Seven Fatality Fire at Remote Wilderness Lodge Grand Marais, Minnesota

Reported by: James W. David

This is Report 055 of the Major Fires Investigation Project conducted by TriData Corporation under contract EMW-90-C-3338 to the United States Fire Administration, Federal Emergency Management Agency.

 FEMA

Department of Homeland Security
United States Fire Administration
National Fire Data Center

U.S. Fire Administration Fire Investigations Program

The U.S. Fire Administration develops reports on selected major fires throughout the country. The fires usually involve multiple deaths or a large loss of property. But the primary criterion for deciding to do a report is whether it will result in significant "lessons learned." In some cases these lessons bring to light new knowledge about fire--the effect of building construction or contents, human behavior in fire, etc. In other cases, the lessons are not new but are serious enough to highlight once again, with yet another fire tragedy report. In some cases, special reports are developed to discuss events, drills, or new technologies which are of interest to the fire service.

The reports are sent to fire magazines and are distributed at National and Regional fire meetings. The International Association of Fire Chiefs assists the USFA in disseminating the findings throughout the fire service. On a continuing basis the reports are available on request from the USFA; announcements of their availability are published widely in fire journals and newsletters.

This body of work provides detailed information on the nature of the fire problem for policymakers who must decide on allocations of resources between fire and other pressing problems, and within the fire service to improve codes and code enforcement, training, public fire education, building technology, and other related areas.

The Fire Administration, which has no regulatory authority, sends an experienced fire investigator into a community after a major incident only after having conferred with the local fire authorities to insure that the assistance and presence of the USFA would be supportive and would in no way interfere with any review of the incident they are themselves conducting. The intent is not to arrive during the event or even immediately after, but rather after the dust settles, so that a complete and objective review of all the important aspects of the incident can be made. Local authorities review the USFA's report while it is in draft. The USFA investigator or team is available to local authorities should they wish to request technical assistance for their own investigation.

This report and its recommendations were developed by USFA staff and by TriData Corporation, Arlington, Virginia, its staff and consultants, who are under contract to assist the USFA in carrying out the Fire Reports Program.

The USFA greatly appreciates the cooperation and information received from Minnesota State Fire Marshal Thomas R. Brace; Deputy State Fire Marshals David Bahma, Terry Christensen, and Glen Bergstrand; and Cook County Sheriff John Lyght. Ms. Mary Nachbar, Supervisor, Public Fire Education and Data Analysis, of the State Fire Marshal's Office reviewed the report and provided additional technical information.

For additional copies of this report write to the U.S. Fire Administration, 16825 South Seton Avenue, Emmitsburg, Maryland 21727. The report is available on the USFA Web site at http://www.usfa.dhs.gov/

U.S. Fire Administration

Mission Statement

As an entity of the Department of Homeland Security, the mission of the USFA is to reduce life and economic losses due to fire and related emergencies, through leadership, advocacy, coordination, and support. We serve the Nation independently, in coordination with other Federal agencies, and in partnership with fire protection and emergency service communities. With a commitment to excellence, we provide public education, training, technology, and data initiatives.

TABLE OF CONTENTS

OVERVIEW 2
SUMMARY OF KEY ISSUES 2
BACKGROUND 3
THE BUILDING 3
THE FIRE 4
CASUALTIES 5
FIRE SPREAD 6
ESCAPE ROUTES AND OCCUPANT ACTIONS 6
CODE IN EFFECT WHEN LODGE WAS BUILT 6
SMOKE DETECTOR PERFORMANCE 6
LESSONS LEARNED 7
APPENDIX A: Pre-fire Photographs of Windigo Lodge 9
APPENDIX B: Windigo Lodge Site Plan 13
APPENDIX C: Windigo Lodge Floor Plans 14
APPENDIX D: Photographs of Fire Scene 17

Seven Fatality Fire at Remote Wilderness Lodge
Grand Marais, Minnesota
July 1991

State Contacts: Thomas R. Brace
　　　　　　　　　State Fire Marshal
　　　　　　　　　Minnesota Department of Public Safety
　　　　　　　　　State Fire Marshal Division
　　　　　　　　　285 Bigelow Building
　　　　　　　　　450 North Syndicate Street
　　　　　　　　　St. Paul, Minnesota 55104

　　　　　　　　　David Bahma
　　　　　　　　　Chief Investigator
　　　　　　　　　Deputy State Fire Marshal
　　　　　　　　　St. Paul Office

　　　　　　　　　Terry Christensen
　　　　　　　　　Fire Investigator
　　　　　　　　　Deputy State Fire Marshal
　　　　　　　　　Grand Rapids, Minnesota

　　　　　　　　　Glen Bergstrand
　　　　　　　　　Fire Safety Inspector
　　　　　　　　　Deputy State Fire Marshal
　　　　　　　　　Duluth, Minnesota

Local Contacts: John Lyght, Cook County Sheriff
　　　　　　　　　Tim Weitz, Deputy Sheriff
　　　　　　　　　David Wirt, Deputy Sheriff
　　　　　　　　　Cook County Sheriff's Department
　　　　　　　　　Grand Marais, Minnesota

　　　　　　　　　Gunflint Trail Lodge Owners
　　　　　　　　　Bruce Kerfoot, Gunflint Lodge
　　　　　　　　　Dana Austin, Rockwood Lodge
　　　　　　　　　Larry Backstrom, Poplar Lake Lodge

OVERVIEW

An early morning fire at the Windigo Lodge in northern Minnesota left seven people dead and six injured. The fire started in the early morning hours of July 12, 1991, as occupants of the lodge lay sleeping. The fire was reported to the Cook County Sheriff's Department from a phone inside the lodge, located at the facility's only enclosed stairway, at 4:21 a.m. There was no organized fire department in the area to respond. Therefore, the Sheriff's Department notified other lodge owners in the area of the fire. The area lodge owners responded to the fire and were able to save two nearby buildings exposed to the fire with hoses and portable pumps. No attempts were made to suppress the lodge fire, since it was fully involved when the area lodge owners arrived. All that was left of the lodge following the fire was ash and metal debris.

SUMMARY OF KEY ISSUES

Issues	Comments
Failure of early warning fire detection	Smoke detectors used for early warning were single station battery operated units. Employees stated the only smoke detector they recalled seeing was on the first floor behind the bar. This would not be sufficient to hear on the second floor had it sounded. State statute requires smoke detector in all sleeping rooms. Smoke detectors were present at the last inspection in 1988; however, none of the survivors stated they heard a smoke detector at any time during the fire.
Absence of a responding fire department in the area	Due to rapid spread of fire and lack of early warning detection or alarm devices, fire department response would have probably made little, if any, difference to the outcome of the fire in terms of life loss. However, damage to the structure may have been less severe.
Failure of exit conditions to provide adequate escape from the interior of the building	Only one of the stairways between the first and second floors offered protection by way of fire-resistive construction. None of the other stairways were enclosed from the first to third floors. All the stairways were open to the corridor system serving the guest rooms. One enclosed stairway did exit directly to the outside; however, it was not a factor in escape.
Automatic sprinklers	None present.
Compartmentation of lodge	Non-existent.
Fire evacuation plans or drills	None.
Outside fire escape	None present.
Code requirements for enclosed stairs	UBC 1973
Fire doors	Guest rooms and at one stairway only.
Interior finish	Drywalls and ceiling with heavy timber construction and exposed beams, with 3/4-inch flame-retardant treated aspen paneling.

Extinguishment finally occurred some ten hours after the fire began by the U.S. Forest Service called by the State Fire Marshal Deputies. This was the first attempt to put water on the fire. It was not until then that fire investigators were able to work the scene.

At the time of the fire, the Windigo Lodge was occupied by 14 people, seven of whom were Windigo employees and seven were guests. The seven who perished in the fire died prior to trying to escape or died while in the process of escaping. Six people were injured in their attempts to escape the building and one occupant escaped without injuries.

BACKGROUND

The Windigo Lodge, on the shore of Poplar Lake, was scenically nestled among towering pines, birch trees, and lovely lakes in the Boundary Waters Wilderness Canoe Area in northern Minnesota just a few miles south of the Canadian border. In 1964, Windigo Lodge was moved from Windigo Point on Seagull Lake to its present location at Poplar Lake.

In November of 1974, the Windigo Lodge caught fire and burned to the ground. In 1975, the lodge was built by the owners, the Ekroot family, and operated until July 1991, when it again burned to the ground.

The Windigo Lodge was licensed by the Minnesota Department of Health for 10 guests on the second floor. However, the night of the fire the lodge was filled to capacity with guests and employees on the second floor, the family and two guests on the third floor (contrary to the license agreement). Also, Mr. Ekroot, who was confined to a wheel chair, slept on the first floor.

This wilderness area, because of its lack of population, has little in fire protection resources. The closest organized fire department is Maple Hill, some 30 miles to the south of Windigo Lodge. To the north of the lodge, approximately 12 miles, is the Gunflint Trail Fire and Rescue Squad; however, it is only equipped to provide emergency medical service (EMS).

There was an informal fire protection effort established by lodge owners in the area. Each lodge owner has a portable pump and a couple of lengths of hose to use for extinguishment. Their purpose is to respond to large fires and protect exposures. The lodge owners have no other firefighting equipment, protective clothing, or self-contained breathing apparatus (SCBA).

THE BUILDING

The Windigo Lodge was a 3-story, wood-frame structure with a partial basement that had a walk-out to the lake. The lodge had fire exposures to the east and west which were single unit rental cabins and a sauna. The south side of the lodge faced Poplar Lake and the north abutted a parking lot and a small liquor store. (See Appendix B, diagram.)

A large, wooden deck wrapped around two sides of the structure nearest the lake. The guest rooms were located on the second floor, which was also used to house employees. The third floor was used to house the Ekroot family and Mrs. Ekroot's mother. The first floor was a bar, dining room, and a lounge area that overlooked the lake. Mr. Ekroot slept on the first floor; he was handicapped and unable to make it up the stairs.

The building was used as a mixed occupancy with assembly areas on the main floor, guest rooms on the second floor, and owner's residence and guest rooms on the third floor. The small half basement was used for storage. The approximate exterior dimensions were 60 feet by 60 feet.

This 3-story structure had no vertical separations. It was constructed in 1975. The exterior walls were partial height concrete block foundation on the lake side and the rest was wood-frame construction with wooden siding. The interior walls were wood frame. (UBC/UFC Type V-N) covered with gypsum wallboard in the guest rooms, corridors, residential areas, and the west stairway. The bar and dining area were covered with 3/4-inch tongue-and-groove aspen paneling and had been flame retardant treated in 1987.

Structural support for the upper floors was provided by exterior walls and interior wooden columns (posts) made of white spruce logs. Interior ceilings were covered on the room side by gypsum wallboard. However, some of the wooden beams which supported the floor joists for the upper floors were exposed. The first floor ceiling was provided with fiberglass insulation presumably for climate control and/or sound transmission. The roof construction is in question; however, it was believed to be rafters shaped in a gable configuration with multiple shed dormers.

The second floor had an interior corridor arranged in roughly a "square" configuration. Ten sleeping rooms were located around the exterior perimeter of the building with solid core self-closing doors between the corridor and the sleeping rooms. In addition to the sleeping rooms, common restrooms were located along the exterior of the second floor wall near the northwest corner. The center "core" of the second floor contained a storage room and one of the stairways leading to the third floor.

Exit Configuration--There were two stairways serving the second floor from the first floor; one located along the west wall and the other located near the northeast corner. The stairway along the west wall was of fire-resistive construction with the base of the stairs forming a vestibule which exited to the exterior or into the bar area through a one-hour fire rated door. The northeast stairway was approximately six feet wide with no headers or partitions to interrupt smoke or fire spread.

There were two stairways from the second to the third floor; one was located along the north wall and the other was located near the center of the structure adjacent to the second floor storage area.

Large casement style windows were present in each of the sleeping rooms. Although not specifically designed as "egress windows," it appears that these were the path of escape used by five of the six survivors.

Fire Protection Features--An automatic extinguishing system was installed in the commercial cooking exhaust system. Single station smoke detectors (believed to be ionization type) had supposedly been installed in the guest rooms, corridors, and at the top of the stairways. Due to the extensive destruction as a result of the fire, no remains of the smoke detectors were found. Although not required by code due to the limited occupant load, exit signs and emergency lighting units had been installed in the corridors of the second floor guest room area. After extensive interviews with guests it was not determined if any of the above were in-place or operated at the time of the fire. Based on observations made at the time of the previous inspection, portable fire extinguishers were distributed throughout the building.

Building Service Features--When originally constructed the building was heated with two unlisted barrel-type wood stoves. Their use was discontinued sometime later, although they may have still been present in the building. At the time of the fire the building was supplied with heat from a wood-burning device located in another building. The heat was piped from one building to another. There was a fireplace in the lounge area, toward the lake side of the lodge; however, its use at the time of the fire was doubtful.

THE FIRE

The fire at the Windigo Lodge occurred sometime between 3:30 a.m. and 4:21 a.m., on July 12, 1991. The lodge was occupied by seven employees and seven guests at the time of the fire.

The fire was reported to the Cook County Sheriff's Department at 4:21 a.m., by Donald Anderson, a guest staying in one of the cabins. Mr. Anderson stated he was awakened by a family member who heard shouting and glass breaking. Mr. Anderson dressed and ran to the lodge where he saw people jumping from windows. He ran to the west side of the building and entered the enclosed stairway vestibule where the phone was located. Mr. Anderson stated it was hot, dark, and smoky and he could not find the phone. He ran from the building to his pickup truck to get a flashlight. He then ran back to the west exit and was able to locate the phone. He was the first person to report the fire.

The Cook County Sheriff's dispatcher immediately reported the fire to a number of neighboring lodge owners. These neighboring lodge owners responded to the Windigo Lodge with portable pumps and hoses. The magnitude of the fire, as described by one lodge owner, was that the lodge building was fully engulfed with flames upon his arrival. Two portable water pumps were placed in-service at the lake, one on the east and the other on the west side of the lodge. These hose streams were used to protect exposures. It was felt there was no sufficient hose streams to extinguish the lodge fire; therefore, no attempt was made.

The Windigo Lodge was allowed to burn throughout the day until the fire investigators were able to enlist the help of the U.S. Forest Service to bring in equipment to extinguish the smoldering remains of the structure. The investigation was then able to begin.

CASUALTIES

The individuals who perished, were injured, or escaped are as follows:

Victim	Floor	Room #	Affiliation	Deceased/Injured/Escaped
Vincent Rolland Ekroot, Sr.	1st	NE corner	Owner	Deceased
Vincent Charles Liestman	2nd	#3	Employee	Deceased
Donald Ray McComb	2nd	#4	Guest	Escaped
Adam Troy Maxwell	2nd	#5	Employee	Injured
Milan Frank Matetich	2nd	#6	Guest	Injured
Michelle Lynn Swenson	2nd	#7	Guest	Deceased
Greg James Swenson	2nd	#7	Guest	Deceased
William Joseph Nelson	2nd	#8	Employee	Injured
Bruce Wayne Kellerhuis	2nd	#9	Guest	Left at 3:30 a.m.
Brian Jay Porter	2nd	#9	Guest	Deceased
Robert William Reed	2nd	#10	Employee	Escaped
Glen Ray Wittman	3rd	West Bedroom	Guest	Injured
Thomas Fredrick Cooley	3rd	Spare Bedroom		Guest
Charlette Geraldine Ekroot	3rd	Master Bedroom	Owner	Injured
Gladys Lillian Merril	3rd	Center West Bedroom	Owner's Mother	Deceased
Duane Anderson	Cabin		Guest	Injured

FIRE SPREAD

Due to the extensive fire damage and the almost complete destruction of all structural members, fire patterns were virtually impossible to observe and physical evidence was almost nonexistent. This lack of physical evidence and observable patterns means that the investigation must center on witness statements and any available fire modeling or empirical data available. The investigation is continuing at the time of this report and the cause has not yet been determined. However, witness accounts suggest that the fire originated on the first floor, possibly in the dining room area. Fire development was fast and overwhelmed occupants as they lay sleeping on all three floors.

It was further determined from witness statements that there was no early warning from smoke detectors; in fact, there is no witness who states they heard a smoke detector at any time during the fire.

The vertical openings in the structure certainly contributed to the rapid spread of the fire from the first floor to the third floor. There was little to stop the spread of fire and toxic smoke. In addition, the lack of early warning contributed to the spread as it may have gone unnoticed for a period of time before discovery.

ESCAPE ROUTES AND OCCUPANT ACTIONS

One of the owners, Vince Ekroot, died in the fire. He was confined to the first floor due to a disability which restricted him to a wheelchair. There was no indication of an attempted escape.

The ten rooms on the second floor, at the time of the fire, were occupied by four lodge employees and five guests. Three of the guests and one employee died in the fire. Two employees and one guest were able to jump from second floor windows, and one employee and one guest were able to escape using the interior building stairway before it became impossible. Two of the three persons who jumped were injured and one of the two using the stairs were injured while attempting to escape.

Four of the five bedrooms on the third floor were occupied at the time of the fire by Charlette Ekroot (owner), Charlette's mother, Gladys Merril, and two guests. Two of the four died in the fire, the other two were seriously injured jumping from the windows.

CODE IN EFFECT WHEN LODGE WAS BUILT

The building code in effect at the time of construction was the 1973 Uniform Building Code. The building code in this area of Minnesota is enforced by the Cook County Building Department.

The Windigo Lodge was the last inspected by the State Fire Marshal Division in 1988 and was on a 3-year inspection cycle. There had reportedly been extensive remodeling done to the building after this inspection.

SMOKE DETECTOR PERFORMANCE

There were reportedly single station, battery-operated smoke detectors on the first and second floors of the lodge. It is uncertain but most likely they were not there. The witness statements said the only smoke detector in the building was located on the first floor near the bar.

LESSONS LEARNED

1. **The choice of early warning detection systems must be adequate for the type of occupancy.**

   The presence and operation of fire alarm equipment in this incident, such as smoke detectors, are somewhat in doubt. The smoke detectors were to have consisted of single station, battery-operated detectors. One witness stated he heard an alarm; however, he believed it to be an alarm clock. Most of the survivors were awakened by other residents.

   The usefulness of alarm devices not electrically powered, electronically supervised, or interconnected for this type of occupancy should be seriously questioned. Single station smoke detectors have value in single family residential settings where the detectors are meant to alert occupants of fire conditions in their immediate area. However, fire alarm systems provide a much higher level of fire safety over single station detectors in this application due to audibility concerns, maintenance, and lack of other fire protection features in the structure. In this particular case there are doubts as to whether the single station detectors were re-installed following the recent remodeling.

2. **Open stairways are almost always a factor in the spread of smoke and fire; alternate, protected exits must be provided.**

   Unprotected vertical openings, three or more stories, were certainly a factor in the spread of fire and loss of life in this fire. Unfortunately, the same openings which allowed the vertical spread of heat and smoke were used as exits and contributed to the delay or impediment of egress from the building.

3. **Code requirements must be complied with; lack of code-required compartmentation contributed to the outcome of this fire.**

   Based on the State Building Code in effect at the time of construction, a one-hour occupancy separation should have been provided between the assembly area (main floor) and the sleeping/residential areas (upper floors) of this structure. This may have provided additional time and provided a protected exit system for the building's occupants had it been coupled with an early warning alarm system.

4. **Evacuation plans and procedures should be practiced by employees and supplied to guests in any hotel, motel, or guest lodge facility.**

   Based on the transient nature of the resort community, having an established plan that is practiced by employees and posted for guests is a vital necessity for this industry.

5. **Decorative elements, such as the exposed ceiling beams in this case, require that special efforts be made to insure the fire-resistive integrity of the construction.**

   It was stated the ceiling construction on the first floor was drywall with exposed beams. However, there were indications from witnesses and photos that the drywall had large gaps where it met at the beams. This allowed fire to travel to the floor construction above and probably contributed significantly to the fire spread.

6. **Facilities serving the public located in rural areas at a distance from fire and EMS services should take extraordinary responsibility themselves for fire and life safety.**

   Dependence upon a fire service organization for immediate fire and life safety protection in this application is not practical. The most efficient and effective fire and life safety protection can only be provided by the facility itself. Ensuring the building is built properly and meets fire and life safety codes, such as early warning detection systems, exiting, enclosure of vertical openings, compartmentation, and perhaps sprinklers, is the beginning for providing protection for guests in this type of occupancy. Further, given the wilderness area in which Windigo Lodge was located, better first aid and firefighting equipment, and the proper maintenance and training of employees in its use, are essential.

   Without the proper precautions for fire and life safety, even if a fire department had been located as close as five minutes from this structure, it is questionable whether it would have been able to positively affect the outcome of this incident.

   Responsibility for one's own safety is also paramount for self protection. Recognizing safe conditions in buildings and not taking the risk of frequenting establishments without solid fire and life safety features would go a long way in ensuring personal survival.

APPENDIX A
Pre-fire Photographs of Windigo Lodge

Windigo Lodge under construction. Poplar Lake can be seen in background at left.

Photo provided by the Minnesota State Fire Marshal

Appendix A (continued)

Dining room and lounge area on first floor. Note ceiling separations at log beams.

Photo provided by the Minnesota State Fire Marshal

Appendix A (continued)

East side of lodge.

Appendix A (continued)

Lake (south) side of lodge showing deck and stairway from basement walkout down to shoreline.

Photo provided by the Minnesota State Fire Marshal

APPENDIX B
Windigo Lodge Site Plan

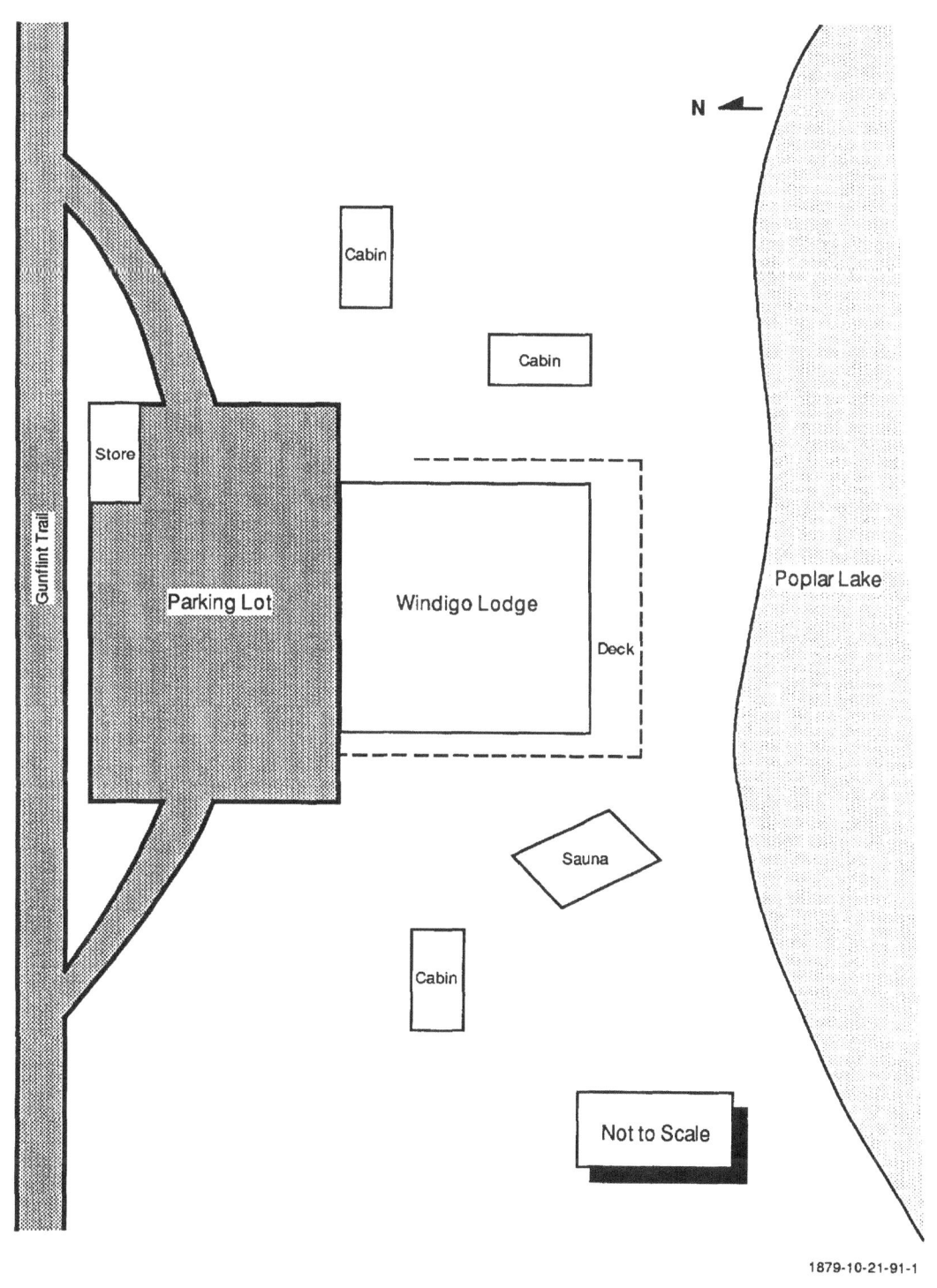

APPENDIX C

Windigo Lodge Floor Plans

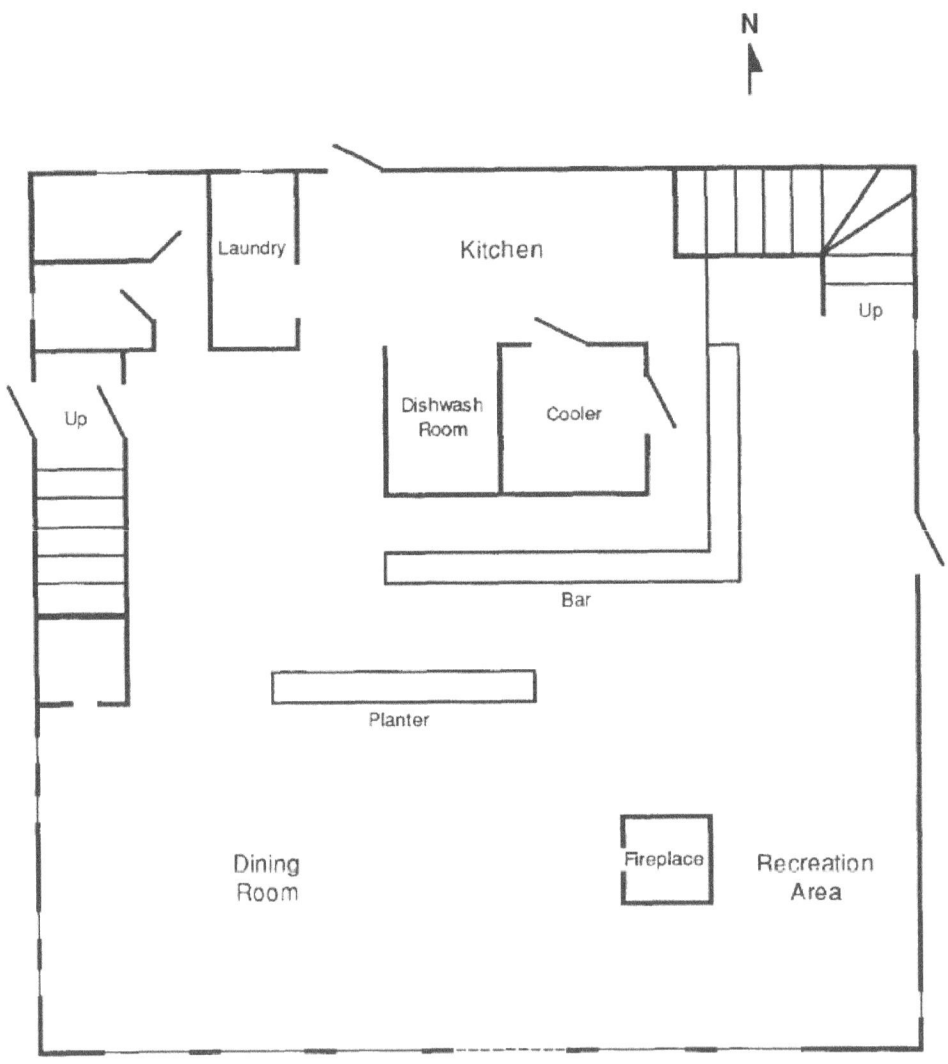

Appendix C (continued)

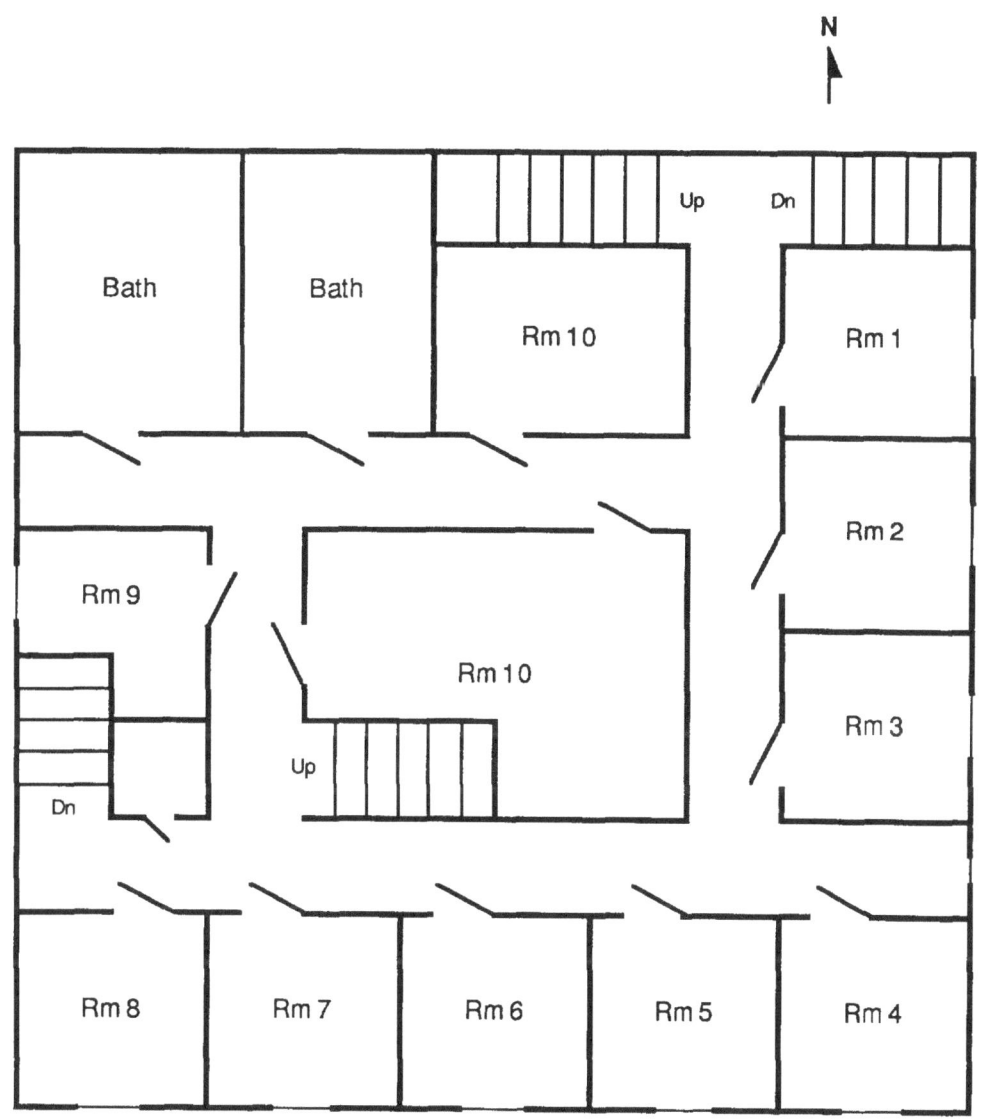

SECOND FLOOR

Not to Scale

Approximate Proportion

Three Story Wood Frame Construction

Appendix C (continued)

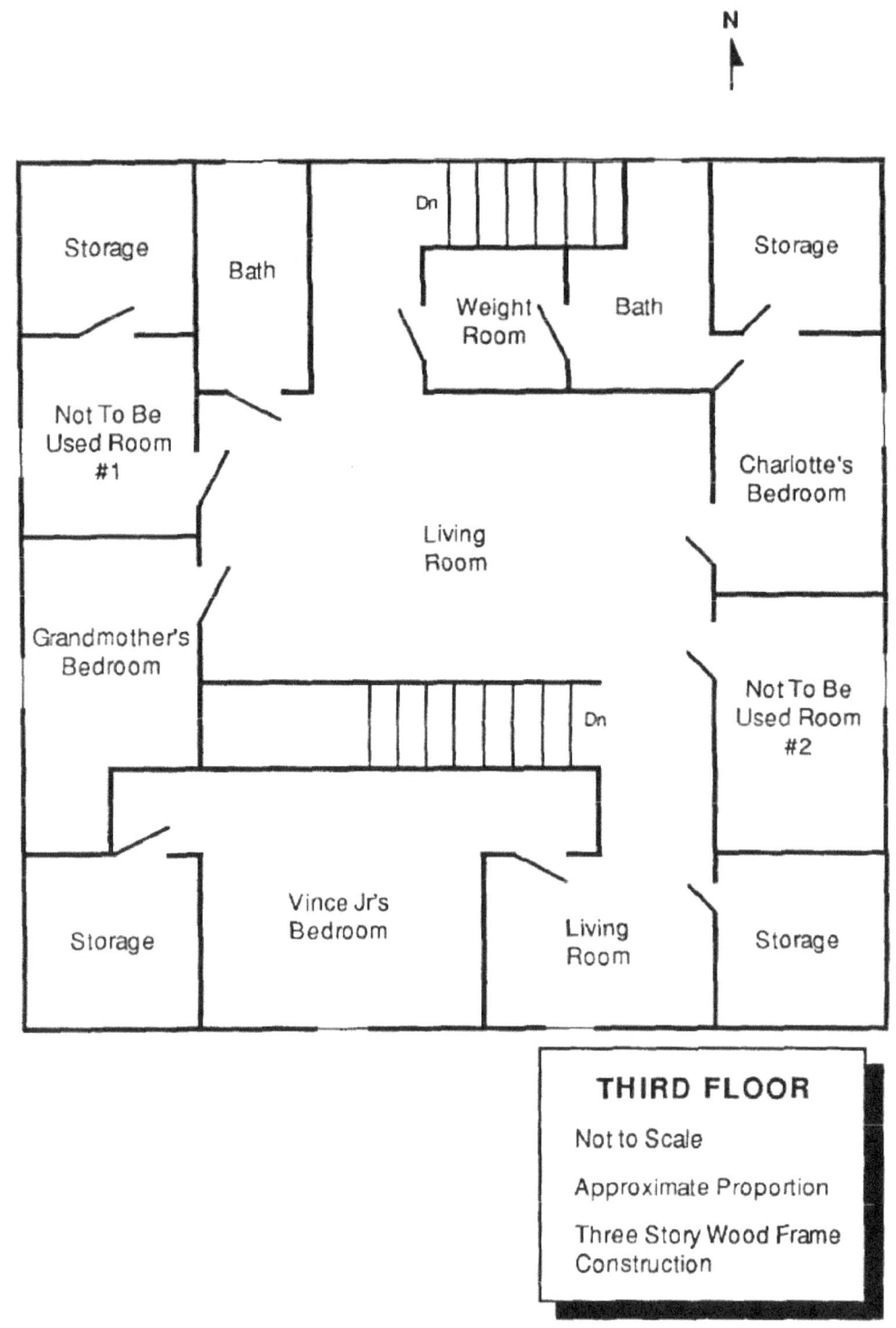

APPENDIX D
Photographs of Fire Scene

View along north foundation wall shows total destruction of lodge by the fire.

Photo by James W. David

Appendix D (continued)

Remains of Lodge from west drive, facing lake.

Appendix D (continued)

Remains of Windigo Lodge from east drive, facing lake.

Photo by James W. David

Appendix D (continued)

Entire structure was consumed by fire; east and north foundation walls shown here.

Photo by James W. David

Appendix D (continued)

East foundation wall and basement walkout to lake (off photo to left); basement wood frame construction consumed by fire.

Photo by James W. David

Appendix D (continued)

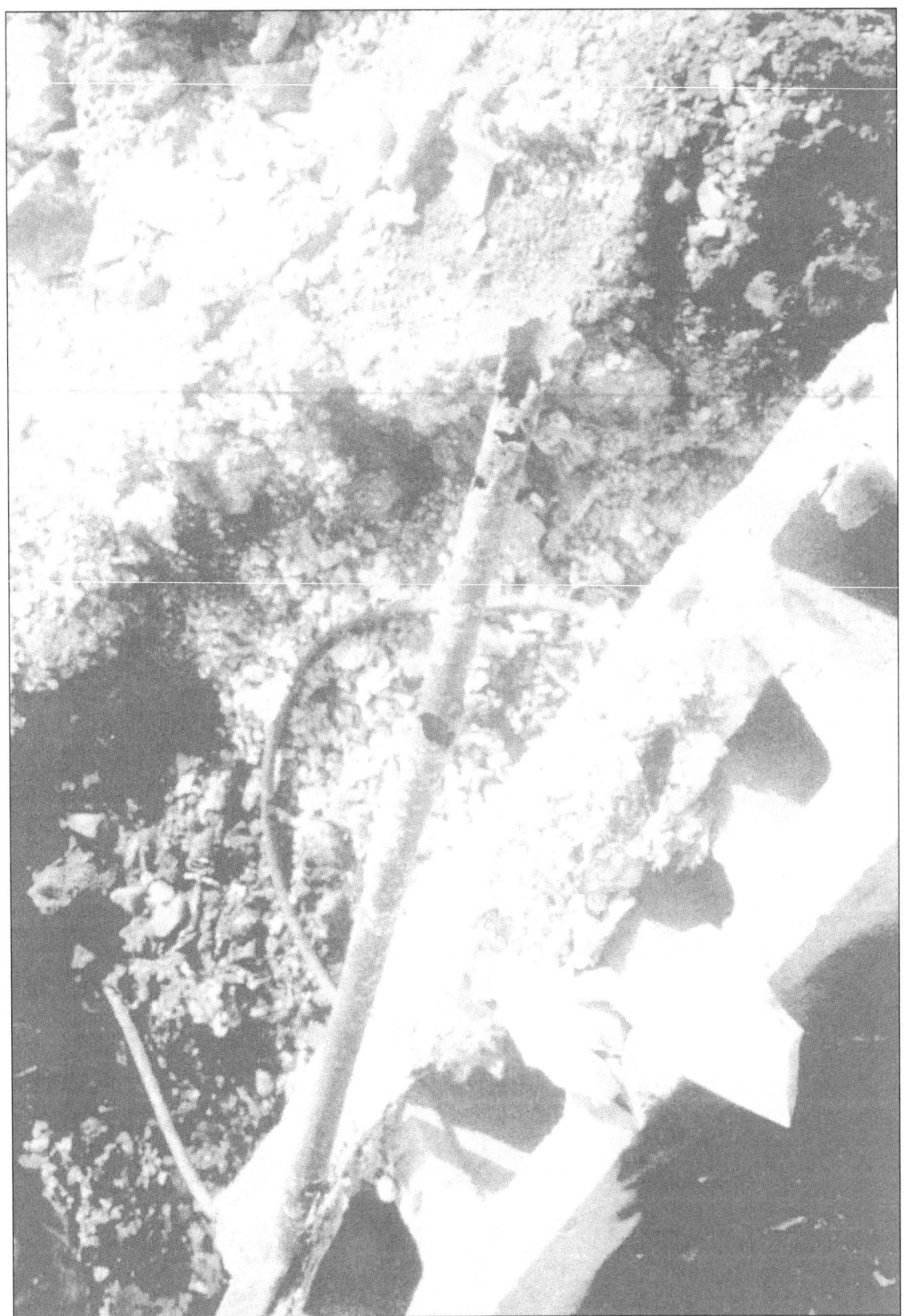

Damage to 3/4-inch copper water pipe from intense heat of the fire.

Photo by James W. David

www.ingramcontent.com/pod-product-compliance
Lightning Source LLC
Chambersburg PA
CBHW081418170526
45166CB00010B/3391